CALIFORNIA FEDERAL
ONE WILSHIRE

LOS ANGELES

This book was devised and produced by Multimedia Publications (UK) Ltd

Editor: Marilyn Inglis
Design: John Strange and Associates
Picture Research: Catherine Blackie, Tessa Paul
Production: Arnon Orbach

ISBN 0 8317 5638 1

First published in the United States of America 1984 by Gallery Books, an imprint of W. H. Smith Publishers Inc., 112 Madison Avenue, New York, NY 10016

Originated by D. S. Colour International Ltd, London.
Typeset by Flowery Typesetters Ltd.
Printed by Sagdos, Milan, Italy.

LOS ANGELES

Leslie Stahr

GALLERY BOOKS
An Imprint of W. H. Smith Publishers Inc.
112 Madison Avenue
New York City 10016

Contents

WESTLAKE
THEATRE

1 The Paradoxical City

Los Angeles is a *déjà vu* city; movies and television have made it so much a part of our collective consciousness that to arrive there for the first time is to return to familiar streets, buildings, canyons, beaches and, above all, freeways.

Three artificial life support systems keep the second largest city in America alive: her freeways, her harbor, and the conduits that carry her water supply from the north. The fourth essential element in L.A.'s existence is the climate.

Los Angeles is a *sunshine* city; in all the volumes that have been written ridiculing and maligning her, the sunshine remains sacrosanct. But the myth that Los Angeles has no seasons, just a boring, perpetual summer, is not true. The signs of her seasons are quite different from those that transplanted New Englanders and Northwesterners are always complaining that they miss, but Angelenos know how to recognize them. Spring is when the succulents, "ground covers" that hold the hillsides in place against wind and rain, bloom in dazzling shades of red and yellow. Fall is when the locals drive out to Calabasas for the Pumpkin Festival, where they stock up on Jack-o'-lantern makings for Hallowe'en. Winter is when they pick tangerines, lemons and avocados off their trees and when the hot, dry Santa Ana winds blow in from the desert, crackling with electricity. Nobody needs to be told when it's summer – although some cynics would say summer in Los Angeles is when ski clothes go on sale (ski slopes are only a two-and-a-half-hour's drive away).

Ironically, the sunshine that attracted the seven-and-a-half million souls who live there is now filtered through the filthy reddish-brown haze of smog created by the autos and industries of those same seekers of the sun. But there are still days, usually after a storm, when the

MacArthur Park is one of 300 in Greater Los Angeles, ranging from vest-pocket size to Griffith Park, more than four thousand acres of ravines, hiking trails, picnic areas, an open-air theater, a 113-acre zoo, a golf course and, on the summit of the hill, the observatory immortalized in *Rebel Without A Cause*.

VonHausen
271-6555
Jeep

Old Mission
San Fernando
Rey de España
1797

mountains, the ocean and everything in between are as clear as they are in the memories of the oldest old timers. To walk barefoot on the beach in the February sunshine and see, in the distance, the snow glistening on the peaks of the San Gabriel Mountains is to understand why they called it the City of the Angels, and to experience the timeless best of what *Los Angeles Times* columnist Jack Smith calls the Big Orange.

Los Angeles is a *haphazard* city; one can't help wondering just what it is that the raft of the city planners actually *do*. Take the case of rapid transit, for example. Until the 1940s, L.A. had an efficient public transportation system – the big red electric streetcars that operated on a network of tracks serving Hollywood, downtown, the Valley and the beaches. The red cars were essential in the city's astonishingly rapid early growth. But in the early 1940s, the city adopted a master plan calling for the expansion of the freeway system to within four miles of every home in the metropolitan area. The red cars were pronounced obsolete, their 1200 miles of track torn up. Today, that right-of-way has vanished under just one more crowded lane of crawling traffic, while officials argue endlessly about how to provide rapid transit and the possibility becomes more prohibitively expensive every year.

As a consequence of this shortsightedness, Los Angeles has become a *car* city. In 1983 there were over five million registered vehicles in Los Angeles County. Daily commutes of a hundred miles and more are not unusual, and popular places such as Disneyland, in neighboring Orange County, are generally considered part of L.A. because, "It's only an hour's drive".

Facing page above The carefully restored Hale house, built in 1895, one of the most cherished old buildings in L.A.

Below Mission San Fernando, 17th in the chain of 21 missions conceived by Father Serra, was built in 1797 by the native Indians.

This page above The circular towers distinguish the Bonaventure Hotel, near the Cloverleaf.

Below Everything seems to bloom just a bit more extravagantly in Beverly Hills – witness these lemon and jacaranda trees.

Los Angeles is a *consumer* city where every imaginable necessity and luxury is mercilessly advertised on television, billboards, radio and in the *Los Angeles Times*, a daily five-pound package of ads (even heavier on Sundays) surrounded by news and features, acknowledged as one of the best papers in America. Despite some pockets of poverty, Los Angeles residents have one of the country's highest levels of disposable income – money to play with – and they like nothing better than to spend it on the accoutrements of the good life: food, fitness, gadgets, entertainment and cars.

Left, above and below The maximum speed on the freeways in L.A. (and anywhere in California) is 55 miles per hour, and traffic laws are strictly enforced. Drunk driving can cost you your driver's license, plus a fine *and* some time in jail. And if the traffic cops don't get you, the Meter Maids will – for not feeding your parking meter.

Los Angeles, an inland city with the most important harbor on the West Coast, is in many ways a *paradoxical* city. Nowhere in America is there a more varied mixture of wealth and poverty, beauty and vulgarity, tolerance and prejudice, generosity and self-centeredness, materialism and mysticism, euphoria and depression, reality and illusion. And in the entertainment capital of the world, it's not always easy to distinguish the reality from the illusion. L.A. has, for example, suffered natural disasters with alarming frequency: fires, floods, mudslides, droughts and earthquakes have destroyed more homes than insurance claims adjusters care to think about. None of those acts of God is real, though, until it happens to you; thus do people continue to build on ashes, sand and fault lines.

Los Angeles is a *commercial* city, accounting for 12 per cent of the country's Gross National Product. More businesses are started here – and more go broke – every year than in any other city.

Entertainment is only one of many major industries, but production companies seem to be everywhere, some as solid as the televison networks, some as insubstantial as the San Andreas fault. Fame and fortunes are fleeting in the business known as show, but the personalities it has attracted have made the city itself a celebrity through the books, movies and television shows they've created.

Facing page above Art critics argue about the merits, or lack of them, of the collection in oil tycoon John Paul Getty's Malibu museum, but none would deny that there's no place like it. This is the late Getty's version of a Roman villa; inside is his richly-endowed collection of statuary, furniture and paintings.

Below The Dorothy Chandler Pavilion, largest and grandest of the three buildings that comprise the downtown Music Center, is the winter home of the L.A. Philharmonic. Visiting opera, ballet and theatrical productions are also seen here, and the Academy Awards are presented on the Pavilion's stage.

This page City Hall Building rises majestically from the surrounding greenery. It stands on North Spring Street and is the center for city administration. If you take a lift to the 27th floor you're guaranteed a fabulous view of Los Angeles.

Los Angeles is an *amnesiac* city; successes and failures are quickly forgotten, and "What's he done lately?" is a common refrain. Above all, though, it's an *enigmatic* city, resisting explanation, wriggling out of definitions. For newcomers, it is, as it has always been, a city of possibilities. it does, after all, refer to its freeways as "arteries". And where there are arteries, there must be a heart.

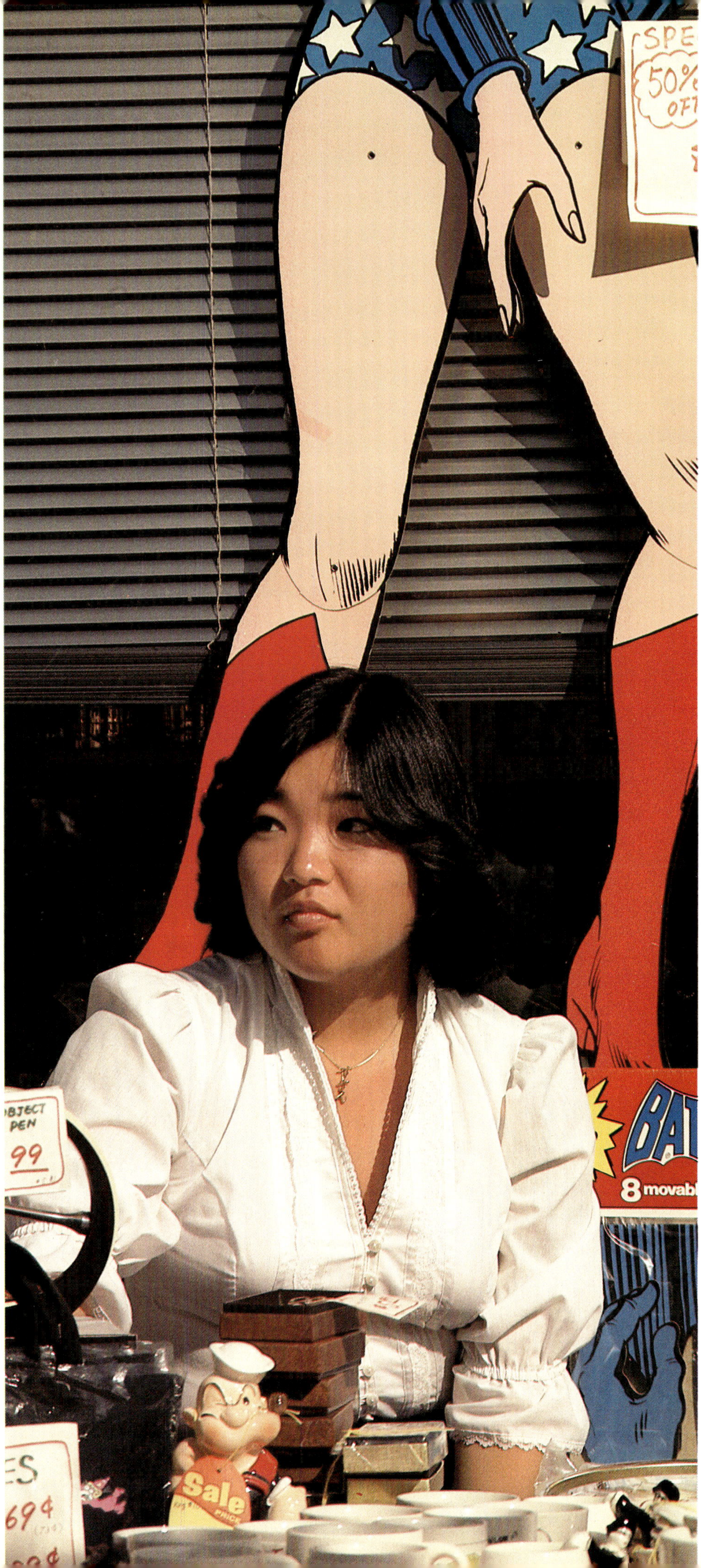

Facing page above These dancers are getting ready to perform at one of L.A.'s many street festivals. Some, like the late lamented Garden Theater Festival, are massive, highly organized, citywide. Others are small, local efforts to raise funds for a charity, show off the local talent, or just have a good time for no particular reason.

Below Los Angeles has something for everyone. It has a pet cemetery. It has psychologists for pets. The city is also rife with gurus, healers, soothsayers, stargazers, spiritualists, numerologers and readers of palms, auras and tarot cards. The Age of Aquarius continues.

This page left The subtlety of the Orient coexists with garish American hype in Little Tokyo, where lovely authentic silk kimonos are sold alongside Batman paraphernalia. Contrasting cultures enrich the city, at the same time creating complex social problems.

Below If you ascend the Palos Verdes cliffs you'll be greeted by this magnificent statue and an expanse of ocean as blue as the sky above. The coves below hide some beautifully quiet and sandy beaches.

2 The Way It Looks

Some cities can be discovered on foot, others by car, but you can only get a true sense of the vastness of Los Angeles from a plane. On a clear day, when the wind has blown the smog away, the view extends all the way from the San Gabriel Mountains to Catalina Island, 26 miles out to sea.

"Greater Los Angeles" is the imprecise but generally accepted term for the 4083-square-mile sprawl that comprises Los Angeles County, namesake of the city that is its seat. Encircled on three sides by the mountains is the "basin" in which L.A. is the largest (at 465 square miles) of 81 incorporated cities between which boundaries are irrelevant except for tax and other official purposes.

The best non-aerial view of the city, and the most romantic, is the 360° panorama seen from Mulholland Drive, the road that runs along the ridge of the Santa Monica Mountains that separate the San Fernando Valley from the rest of the city. Between Thanksgiving and New Year, thoroughfares all the way from Hollywood Boulevard, at the foot of the hills, to the harbor twinkle with colored holiday lights. No wonder Mulholland Drive has been L.A.'s most popular lovers' lane ever since the first Model T Ford chugged up one of the canyons!

From the vantage point of Mulholland you can see that the city so often described as having no center does indeed have a geographical, if not spiritual center: the four-level downtown freeway interchange, nicknamed for its "Cloverleaf" shape. The Cloverleaf, where more than a million cars crawl in bumper-to-bumper rush hour traffic every day, is a necessary evil for L.A.'s 3.3 million commuters. If the collective stress experienced by those drivers could be harnessed, it would solve the energy crisis!

L.A.'s skyline has changed dramatically since the days when earthquake codes prohibited highrise. Now, skyscrapers are built to bend, not break, when the San Andreas fault shifts enough to cause the 'big one' that scientists agree is inevitable. What they can't agree on is when. The last major jolt was in 1971, in Sylmar, at the northern edge of the San Fernando Valley.

First Interstate
ONE WILSHIRE
ROBINSON'S
520

This page above, below and right One-hundred-mile-a-day commutes are common on L.A.'s 725 miles of freeway; only oddballs question the sanity of spending up to 15 hours a week cursing and sweating in freeway traffic. And even though the speed limit is a conservative 55 miles per hour, negotiating the freeways takes a special kind of nerve. The timid and the phobic use surface streets which are, statistically, no safer – but sometimes faster!

Because Los Angeles likes to tear down her old buildings (instead of following the example of her rival city, San Francisco, and restoring them), the complex of handsome commercial and residential highrise buildings surrounding the Cloverleaf is comparatively new. Most prominent are the World Trade Center, the Bunker Hill Towers, the 51-story Arco Towers and the swanky 35-story Bonaventure Hotel, with its five glistening circular towers. All are designed to bend, not break, in an earthquake. In the shadow of these skyscrapers, and more famous, since the 1984 Olympics, than all of them put together is the Memorial Coliseum, a faithful copy of the Roman prototype.

Adjacent to the Coliseum is Exposition Park, with its outstanding rose gardens – a welcome green spot in downtown L.A., where the dominant elements are concrete, steel and glass. Even as the business district bustles and prospers, the streets that surround it are depressing and squalid. Derelicts, many of them former and potential mental patients, congregate in and around the Greyhound Bus Station, the Public Library, and the saloons and soup kitchens of Main Street, also known as Skid Row. Some parts of Los Angeles look a lot more attractive seen from Mulholland than close up.

Just east of the Cloverleaf the streets are crowded so densely with Latinos that you wonder whether you've been magically transported to Tijuana, the impoverished border town just a hundred miles to the south. This is the edge of East Los Angeles or the *barrio*, the Spanish word for neighborhood. More than two million Mexican-Americans live in Los Angeles, many of them in the *barrio*. Except for the touristy Olvera Street, the *barrio* is as unfamiliar to Angelenos not of Mexican descent as the Arctic. To them it's a place you pass on the freeway, on the way to Disneyland. Also on the way to Disneyland are the City of Industry and the City of Commerce, both as ugly as their names – flat, smoggy, dusty wastelands made profitable by businessmen who build warehouses and factories on the cheap land.

Right A car is a necessity in Los Angeles. Driver's Education is a required high school course and acquiring your first car, usually at the age of 16, is an almost-obligatory rite of passage. On Saturday nights, Hollywood and Van Nuys are jammed with 'cruising' teenagers; without wheels, a date is unthinkable.

South Central Los Angeles is a euphemism for Watts, the 2.5-square-mile black ghetto that nearly destroyed itself with fire and looting in the riots of 1965. With its broad, tree-lined streets and frame bungalows with front and back yards, Watts couldn't look less like the stereotypical slum. But in the midst of this bleakness rises one of L.A.'s architectural wonders: three towering cement structures, two of them ten stories high, densely decorated with bits and pieces of metal, sea shells, broken tiles and bottles and dishes. These are the Watts Towers, Simon Rodia's monument to all those who walk, skip or dance to the sound of a different drummer.

At Los Angeles Harbor, the largest manmade port in the world, the landscape is dominated by cargo ships, passenger liners, fishing boats, shipyards, warehouses, storage tanks and canneries. The harbor is never silent; horns, whistles, the grinding and clanging of metal on metal accompanies the action that continues around the clock.

The brighter, lighter face of L.A. lies to the north and west of the Cloverleaf, where the beautiful people live, play and spend, if not earn, their money. Just north of the freeway interchange are Dodger Stadium and the hilly 525-acre Elysian Park, where hippies staged massive love-ins in the 60s. Not far away is Barnsdall Park, with Frank Lloyd Wright's famous Hollyhock House (built in 1917) and the Municipal Arts Gallery. Still further north, in the foothills, is the largest park within the boundaries of any American city, Griffith Park.

Two freeways cut directly through the hills from the city to the San Fernando Valley, but the scenic routes are those that wind through the canyons: Topanga, Coldwater, Benedict, Nichols, Outpost, Laurel, Beverly Glen.

The far side of the canyons is a sea of bedrooms: the 219-square-mile San Fernando Valley, in which more than a million Angelenos make their home. Before World War II, the Valley consisted of orchards of fruit trees, groves of walnut trees, fields of corn, and grazing land for horses and cattle. Postwar prosperity brought a boom in the construction of

This page above Wilshire Boulevard is as long as Sunset, but considerably more conservative.

Below Angelenos live with smog the way some people live with chronic diseases.

Facing page City Hall.

UNION BANK
TWA
WESTLAKE
THEATRE
WORLD
SAVINGS
Crown Royal
WORLD SAVINGS

Left The skyline as seen from MacArthur Park, a refreshing green spot surrounded by the steel, glass and concrete of downtown Los Angeles. Exposition Park, adjacent to the Memorial Coliseum, is another downtown oasis, especially when the roses are in bloom.

Below This has to be trick photography; the freeways are only this empty at 4 a.m. The Union Bank Building on the left and the World Trade Center on the right, both built in the 60s, are a generation younger than the Southern California Gas Company building in the center.

LIMITED
HOT
HAMBURGERS
DOGS
UNION PACIFIC
535
OPEN
PARK
PROBABLY
THE BEST
HOT DOGS &
HAMBURGERS
IN THE WORL

420 PQA

housing developments – acres of one-story single-family houses, many with swimming pools, three-car garages and air conditioning, the latter a concession to the 100°-plus summer temperatures. After the "tracts" of houses came the freeways, followed by apartment complexes, shopping malls, the Magic Mountain amusement park and the Busch Bird Sanctuary. Two historic sites survived all this development; they are links in the chain of 21 missions founded throughout California by Father Junipero Serra. San Fernando Rey de España dates back to 1797; the San Gabriel Mission is even older. Founded in 1771, ten years before Los Angeles was settled, it is the third oldest structure in California.

The Valley is lush with lawns, flower and vegetable gardens, trees laden with apricots, tangerines, lemons and other fruits; only the proliferation of cacti and succulent plants reminds you that this land is semi-desert, reclaimed by billions of gallons of water piped in – some say stolen – from Colorado and Northern California.

On a swelteringly hot summer day, there are two welcome oases of cool tranquility in the Valley. Descanso Gardens, in La Canada, has one of the world's largest collections of camellias – more than 100 000 plants, blooming on 165 acres of oaks, streams and hillsides. The botanical gardens that frame the Huntingdon Art Gallery and Library in San Marino have no equal in California, and few anywhere in the world. The art gallery's most popular attractions are Gainsborough's *Pinky* and *Blueboy*; the library, a boon to scholarly researchers, contains such priceless manuscripts as a 15th century Gutenberg Bible and Shakespeare's First Folio. On the 207-acre grounds, Angelenos and international visitors enjoy the magnificent desert garden, the rose garden (containing over 1200 varieties, some dating back to before Christ), and the herb, Shakespeare, Japanese, Australian, Zen, camellia and jungle gardens.

Facing page above Eating out is a way of life in Los Angeles, whether in wildly expensive expense-account restaurants such as Ma Maison and Perino's or the fast food operations that offer everything 'to go'.

Below No, this isn't a set for *Star Trek* – it's a restaurant in the center of the parking lot at Los Angeles International Airport, familiarly known as LAX.

This page Olvera Street, the oldest street in Los Angeles, is on the edge of the *barrio*.

The
Beverly
Hill

Facing page The Beverly Hills Hotel – symbol of Hollywood's most glamorous, most luxurious present and past. The grounds are lush, the service incomparable, the rooms exceedingly private (some are in detached bungalows), and the famous Polo Lounge is still the scene of make-and-break lunches.

This page left This Rodeo Drive shopping center is in Beverly Hills, in the height of the high rent district. Within just a few blocks of the intersection of Rodeo Drive and Wilshire Boulevard are I Magnin, Saks Fifth Avenue, Tiffany's, Bonwit Teller's, Cartier's, Van Cleef & Arpels, Mr. Guy, Hermes, Courrèges and Le Dernier Cri.

Below Los Angeles, the consumer's heaven, has shopping malls galore, all expensively air-conditioned even when the weather outside is perfect. The glamorous new Beverly Center Mall on La Cienega Boulevard contains no fewer than 14 movie theaters!

Below This mural was painted as part of a public jobs program to provide training and work for the unskilled or unemployed. Most such programs were discontinued when the taxes that supported them were cut, let us hope that this muralist has gone on to better, if not bigger, things.

Facing page above A detail of the Watts Towers, the realized dream of Simon Rodia, an Italian immigrant tile setter. At the age of 40, Rodia began building the towers in his back yard. His neighbors called him the junk man; they also called him crazy. Thirty-three years later, in 1954, Rodia completed what is now acknowledged as a masterpiece of 20th century folk art and moved away.

Below Nearly a million blacks live in Los Angeles, many of them in Watts.

What makes Los Angeles unique among cities lies neither to the north, south nor east of its center but, fittingly, to the west, the direction synonymous with adventure. Driving west from the Cloverleaf, not on the Hollywood Freeway but on Sunset Boulevard, you come to the hilly, picturesque community of Silver Lake, named for the large round pond in its center. Beyond it is Hollywood, of interest only to tourists, sociologists and film historians. A few blocks farther west is the trendy Sunset Strip. It's a mishmash of high-rent offices, high-priced shops, glitzy billboards, and restaurants where probably more deals are discussed (and fewer consummated) than anywhere else in town.

The Strip leads into Beverly Hills, a city within a city; it has its own police department, school system and city hall.

This page above The past twenty years have brought a flood of refugees from Southeast Asia.

Below Both Chinatown and nearby Little Tokyo have some fine restaurants, but don't judge the food by the decor; some of the best meals are served in the most unimpressive surroundings.

Facing page above The Old Plaza is the city's birthplace. Two of the oldest buildings, the Old Mission Church and the nearby Avila Adobe, both built in 1818, have miraculously survived earthquakes and bulldozers.

Below Tacos, tostadas and enchiladas are so common in L.A. that they're no longer considered foreign foods.

El Acapulco: MARISCOS
MEXICAN FOOD
TACOS CARNITAS • ASADA
MENUDO
FOOD TO GO
PARKING IN REAR
El Acapulco Mariscos
MEXICAN FOOD
TACOS CARNITAS ASADA
MENUDO
FOOD TO GO
PARKING IN REAR
NUEVO DUEÑO
MARISCOS
TACOS CARNITAS ASADA
MENUDO
FOOD TO GO
COMIDA MEXICANA Y SALVADOREÑA
BIRRIA, B.B.Q, Y LONGANIZA x LIBRA

3 The Way They Live

L.A. is a city outsiders love to hate – witness the unflattering portraits of it in innumerable novels, including those of Raymond Chandler, F. Scott Fitzgerald's *The Last Tycoon*, Aldous Huxley's *After Many A Summer Dies the Swan*, Evelyn Waugh's *The Loved One* and Nathaniel West's *The Day of the Locust*. Its "anything goes" attitude – its easy enthusiasms, short attention span and obsession with novelty inspire jealousy and make it an easy target. Angelenos are free to be as outrageous as they like; for better and worse, "Do your own thing" is their motto.

Contrary to a common misconception, the film industry does not support Los Angeles. Hollywood has a high profile but it is only one small part of L.A., and the closest most Angelenos have been to the movie business is the Universal City tour. The Hollywood types on which Fitzgerald and other novelists based their characters still exist, but the Hollywood community is hardly typical of the L.A. lifestyle. In fact, there are far too many L.A. lifestyles to enumerate. One reason for this is the incredibly varied ethnic mix of the inhabitants. The film industry "discovered" Los Angeles in 1910, but it wasn't until the 1940s that the city's phenomenal growth began in earnest. Aircraft workers who came west for jobs stayed on after the War; so did soldiers and sailors on their way back from the South Pacific. There had always been a trickle of immigrants from Europe, Hawaii and the Far East; since the wars in Southeast Asia, the trickle of Orientals has become a flood. There are now sizeable neighborhoods of Vietnamese, Thai, Korean and Cambodian refugees, as well as significant numbers of Puerto Ricans, Salvadorans, Cubans, Samoans, Lebanese, Arabs, Armenians and Filipinos.

Officially, over two million Mexicans live in Los Angeles; there are perhaps

Angelenos do their own thing with a vengeance, for better and worse; anything goes in L.A., and some of the characters you see on the streets are as bizarre as any you'll see in the movies.

another million who cross the border illegally in order to earn money with which they support families back home. At the extremes of the color spectrum are the Anglos (now a minority) and nearly a million blacks, most of whom are descended from immigrants from the rural South. This rainbow of races enriches the city's culture but creates mammoth social problems; at least 85 languages are spoken by children in the L.A. School District, where 120 000 students have what is termed "limited English proficiency".

One of the oldest ethnic neighborhoods in L.A. is the Fairfax district, a section of West Hollywood where Jews, mostly from Eastern Europe, settled before World War II. Their children have long since moved to the more heterogeneous suburbs, but the elderly Jews remain, unwilling or unable to start over still again. Cantor's, the most authentic Jewish delicatessen in town, is a landmark, its bagels and blintzes a tradition. Like many L.A. restaurants, grocery stores and gas stations, Cantor's is open 24 hours a day.

An excellent place to indulge in the cheap entertainment of peoplewatching is on Fairfax Avenue at Third Street: the Farmer's Market, where the people tend to be as exotic as the food. Other fruitful peoplewatching locations are the Venice

Facing page above Gold's Gym, a national mecca for body builders, is very near this popular outdoor workout spot just off the Venice Boardwalk.

Below In their shiny black wetsuits and swimfins, surfers look like huge mutant crows, sitting patiently on their boards, waiting for the wave like this one that will make the day worthwhile. Sometimes they're joined by an inquisitive sea lion. In the spring, they can watch the procession of California gray whales that migrate from Baja California to Alaska.

This page left Peoplewatching is a favored pastime at the beach.

Below Skaters now outnumber cyclists on the bike path in Hermosa Beach by about five to one. The portion of the Palos Verdes Peninsula in the background is called Hollywood Riviera, in honor of the European resort to which it bears – topographically, at least – a striking resemblance.

LA

boardwalk, any of the fishing piers, Hollywood Boulevard, the boutiques of Silver Lake and Melrose Avenue, the Beverly Center Shopping Mall, and Westwood on a Saturday night. Except for Silver Lake, all these locations are on the west side. So is Beverly Hills, also good for peoplewatching – but walking is so rare there that a pedestrian outside the shopping area may be stopped by police, suspected of burglary or worse.

L.A.'s mild climate – the mean temperature is 64° – means that dress, like behavior is casual. During most months bare legs, arms and heads are ubiquitous, although some restaurants have signs reading SHIRTS AND SHOES REQUIRED. Even in the hottest weather, humidity is low and nights comfortably cool, conducive to backyard barbeques and outdoor concerts in the Valley and at the Hollywood Bowl, in the hills just off the Hollywood freeway.

In a city built on a web of geological fault lines, it's not surprising that the emphasis is on style over substances. Los Angeles sometimes appears to be a huge market where youth and beauty, or the illusion of them, are always on sale, seductively advertised in newspapers and magazines and on billboards. Shopping is a popular pastime. The rich can shop in Beverly Hills or on Wilshire Boulevard; bargain hunters and treasure seekers have plenty of neighborhood swap meets from which to choose. A gargantuan swap meet is held monthly at the Rose Bowl, better known for its championship collegiate football games on New Year's Day.

Eating out is a way of life in Los Angeles. Like most Americans, Angelenos tend to be confirmed carnivores. Some of their favorite restaurants serve only steak, prime rib,

Spectator sports are wildly popular, and not only on television. These are baseball fans at Dodger Stadium. Fans also flock to the Coliseum for football, the Sports Arena for basketball and soccer, the Forum for hockey, Hollywood Park for horse racing and to Gardena's Ascot Park for auto racing.

This page above This airborne skateboarder, whose board seems nailed to his shoes, is perfecting his skill at a skateboarding rink in Redondo Beach. Helmets and kneepads are required here only because the management is wary of lawsuits.

Below These East L.A. gang members look pleasant and friendly, but street gangs are a major cause of crime in Los Angeles.

This page right A shoe shine stall.

Facing page above The Los Angeles Police Department tries various ways to improve its relations with 'minority' citizens.

Below A blank wall doesn't stay blank for very long in Los Angeles – what the muralist don't decorate, graffiti 'artists' deface.

V13
Gold
Disco
Funk
Punk Rules
Germs
Rock & Roll will
Never Die!!!
Punk
OUT
Venice
Punks
REED
SUNSET SALOON

LOS ANGELES
1984
1984

BABY POWDER

Facing page above Festivals take place frequently in East Los Angeles, the Mexican-American *barrio* – the Blessing of the Animals, held on the day before Easter, is one of the most popular festivals. Cinco de Mayo (the Fifth of May) is the most colorful.

Below This children's barbershop is unusual in Los Angeles in that the customers are not trying to look any younger than they are.

This page left A rainbow of faces enriches the city's culture – at least 85 languages are spoken by children in the Los Angeles School District. At Hollywood High School alone, students speak 38 languages.

Below The boardwalk at Venice Beach, once the preserve of pedestrians and cyclists, was taken over by skateboarders in the 60s and by rollers skaters in the 70s. This cyclist has invented an ingenious contraption in which he can nap while he waits for the skaters to get out of his way.

barbecued ribs, or charcoal-grilled hamburgers and hot dogs. But they also love their veggies, and Los Angeles, the birth-place of the salad bar, is unequalled in the quality and variety of her vegetables and fruits.

Angelenos patronize drive-in restaurants, movies, banks, laundries and churches. They spend nearly as much time washing their cars – another ritual – as watering their lawns. They strive to outdo each other with personalized license plates that can be funny (I'M OLD) or philosophical (IBI UBU) or boastful (RICH) or sentimental (4 MOM) or teasing (TRY ME) or a simple statement of the owner's name or nickname, in case he or she should forget it. If cities had symbols, as states have flowers and people birthstones, the most appropriate one for L.A. would be a gas pump.

A car can mean freedom, but there are also disadvantages to a car culture. Smog is the most obvious; aside from the aesthetics of looking up at a brown rather than blue sky, there are days when schoolchildren are not allowed to play outside because of the health hazard, and anyone with a delicate respiratory system would be more comfortable in other climes. In spite of restrictions on industrial pollutants and auto exhaust systems, the smog persists, killing trees all the way up to the 5000-foot mark in the San Gabriel and San Bernardino mountains.

Another disadvantage is loneliness. Any big city can be a chillingly lonely place, but L.A.'s automotive lifestyle can make it lonelier than most. Spending one-eighth or more of your waking hours alone in an automobile is bound to create feelings of isolation, and the pages of L.A.'s weekly papers are filled with "Personals" seeking possible antidotes to loneliness: love, marriage companionship and, of course, sex of every description.

In the sections of town where the action is – Beverly Hills; Westwood, West L.A., Century City, downtown and especially at the beach, everyone is either young or desperately trying to look at least ten years younger than their age. And the beauty they cultivate isn't just skin-deep. "Personal growth" is a growth industry in Los Angeles, and alongside those Personals ads for Mr., Ms. and Mrs. Right are notices of dozens of self-improvement classes, therapy groups and workshops. There are nearly 2000 churches in L.A., some belonging to sects found nowhere else.

Death is unwelcome to all of us, but

Facing page above On Saturdays and Sundays, roller skaters on the Venice boardwalk vie to see who has the most outrageous costume and the best dance routine.

Center Murals are on view in some unexpected places in L.A. – this one is just around the corner from the scene it depicts on the Venice Boardwalk.

Below Los Angeles has its share of senior citizens, but they are conspicuous by their absence just about everywhere except their own neighborhoods.

This page left What makes this picture special is that it was taken in February, the day after a rain storm. The sun shines more than 300 days of the year in Los Angeles, and residents make the most of it. Rain is so rare that most of them don't own an umbrella or raincoat.

Below Vigilant against any perceived threat to their freedom to die as they wish, Angelenos have consistently stopped legislation that would require them to wear motorcycle helmets or seat belts.

This page left This skater will perform his carefully choreographed dance on the Venice Boardwalk to the sound provided by his battery-operated tape recorder. However elaborate their routines are, skaters never ask onlookers for donations. They perform only for love never for money.

Below This enterprising man's 'shop' consists of two shopping carts at a downtown intersection.

Facing page above Shopping is a popular pastime for Angelenos; the most elegant shops, such as the Rodeo Collection shown here, are in Beverly Hills.

Below The Farmer's Market is an authentic piece of Hollywood's history. Its low whitewashed frame buildings haven't changed since the days when small independent growers displayed their freshest, finest produce, and stars really did eat at its outdoor cafés.

particularly so to Angelenos; given the choice, they'd rather see the movie than die. Since they haven't the choice, they have created the only cemetry in the world that doubles as a major tourist attraction: Forest Lawn. Out-of-town visitors are often taken to see this Hollywood version of death, where Forest Lawn coloring books and maps of 141 celebrities' graves are for sale in the gift shop.

More conventional tourist attractions are Universal Studios, Knotts Berry Farm, Magic Mountain Amusement Park, Lion Country Safari, Movieland Wax Museum, Chinatown, Little Tokyo, Farmer's Market, Marineland and of course the most popular of all, Disneyland. All are well worth seeing, but none more so than certain "hidden" attractions known only to the natives, who cherish them for the beauty and relative obscurity. This private, tranquil Los Angeles is far removed from the city's public image as a wild and crazy place, and not even its staunchest defenders would deny that the wildness and craziness outweigh the tranquility.

In page upon page of listings, local papers and magazines tell of entertainments to delight the eye, ear, taste buds and mind. The Comedy Stores in Westwood and Hollywood give youngsters who dream of becoming the next Richie Pryor or Joan Rivers a place to try out material. A dozen jazz clubs present first class performers, even though some clubs open and close with alarming speed. Movie theaters cater to all tastes, specializing in everything from silents and revivals to the latest, most far-out foreign imports. Filmex, Hollywood's annual film festival, is going strong as it approaches its 15th year. Movies are as popular as ever – but not, sad to say, in the theaters. Increasingly, people prefer the convenience of watching videos at home to going out to see a movie as part of an audience. An era ends; an era begins.

This page left The first day in nearly four years when a bank was *not* robbed in Los Angeles was just before the 1984 Olympics, when police manpower was deployed at 100 per cent of its strength.

Below Despite predictions of dire catastrophes, from hopelessly snarled traffic to outright acts of terrorism, the 1984 Olympics turned out to be a triumph of timing and spectacle.

Facing page above The 1984 Olympics were the biggest event to hit Los Angeles since – well, since the 1932 Olympics, also held here.

Below An empty Los Angeles Coliseum.

NISSAN
NISSAN
Coca-Cola
CAMEL
Welcome to Miller Time

4 The Coast

The beach looks its best at dawn, after the tide has gone out, taking with it yesterday's footprints and leaving a slate as clean, or so it seems, as on the day it was created.

If it's after Labor Day but before Memorial Day, the beach will remain relatively empty all day. This is when those who love it love it most of all. But if it's summer, or even a warm day in the spring or fall, it will soon be crowded with people of all ages and descriptions from all over the county. The farther they've traveled, the more they've brought with them. If it's a real scorcher of a day, the beach will be wall-to-wall towels, the lifeguards will be busy hauling people to safety from the treacherous rip tides that can deceive even experienced swimmers.

All beach communities are the same; all beach communities are different. The natural elements – sounds, smells, sand and sea – are eternal, albeit somewhat polluted; but the structures on the shore define the differences between, say, the Malibu Colony and Muscle Beach.

Nowhere is the altered character of the coastline more evident than in Venice, founded at the turn of the century by a tycoon who had fallen in love with the Italian city and tried to duplicate it. The canals and the gracefully curved bridges remain, but gone are the gondolas, the luxurious St. Mark's Hotel (at which Sarah Bernhardt once stayed), and most of the original "palazzos". Today, the owners of the quarter-million-dollar (and up) condominiums overlooking the canals pay more for a month's parking than their predecessors did for a month's rent, and only the rich (and the ducks) can afford to live on the Grand Canal.

The rich who choose not to live at the beach may wish to keep a boat there instead. Thousands of pleasure craft are moored at marinas from Marina del Rey (the largest) to Los Angeles Harbor's Cerritos Channel (the busiest), and the demand for slips is so great that some waiting lists are two years long.

It isn't easy, but it is possible to find solitary beaches within the borders of the city. Those who are resourceful and determined enough to scramble down the Palos Verdes cliffs may be rewarded with an isolated cove like this one where, on summer nights, you can see the grunion run.

Facing page above These boats, part of the largest commercial fishing fleet in the U.S., are tied up at Los Angeles Harbor, the fifth busiest in the country and the largest man made port in the world. Through this gargantuan industrial enterprise pass 26 million tons of raw and manufactured goods a year.

Below The lucky few who live within walking distance of the beach have a proprietary feeling about it. To people who have grown up on or near the beach, it's inconceivable that anyone might never have actually bathed in an ocean.

This page left Hang gliding has become very popular.

Below With an average rain fall of only 15 inches a year, and most of that in the winter months, joggers are everywhere except on the freeways.

This page Hopeful fishermen (and women) begin casting their lines well before dawn from the Santa Monica pier. On a good morning they'll take home perch, bonita and shark. On a bad day, the catch may be a jellyfish and an old boot. The famous Santa Monica carousel (scenes in *The Sting* were shot there) is on this pier, and still in operation.

Facing page above When the public objected to the ugly oil derricks offshore, the oil companies created this ingenious disguise.

Below In the summertime, Will Rogers State Beach, at Santa Monica, is covered with wall-to-wall beach towels and sun worshippers. The Santa Monica Mountains in the distance are a paradise for hikers and nature lovers, especially when the spring wildflowers are in bloom.

5
Hollywood!

Like so many of its residents, Hollywood is unmistakably a has-been – its famous boulevard largely a jumble of tacky restaurants, massage parlors, bars, and shops selling pornography, kinky lingerie and tourist souvenirs. The traffic is horrendous. At Hollywood and Vine, the most famous intersection in the world – and possibly the most anti-climactic, with its ordinary office buildings and geriatric department store – the "Walk of Fame" begins. Unfortunately, the Chamber of Commerce, which dreamed up the project, didn't consider that sidewalks are, inevitably, filthy, and that no star can shine under a constant coat of grime.

Still, a few parts of the boulevard manage to shine with faint reflections of former glory. The Santa Claus Lane Parade still takes place on Thanksgiving night, climaxed by the float bearing the star (more likely of television than of movies) who is the year's Grand Marshal. C.C. Brown's, the ice cream parlor that raised the hot fudge sundae to an art form, still has its stylish polished mahogany booths, marble-topped tables, "Casablanca"-type ceiling fans and flocked wallpaper, and still serves its speciality in a silver goblet with the bubbling fudge in a pitcher on the side. Three long-established restaurants have managed to survive with their high standards intact: Musso & Frank's, Don the Beachcomber's and The Brown Derby (the latter two just off the boulevard).

The area eight miles northwest of the central city was christened Hollywood in 1896 by an early settler, Mrs. Daieda Hartell Wilcox. Mrs. Wilcox had left the Midwest with a mission: to establish a God-fearing, teetotaling community in the sunshine near the sea. She and her husband bought and subdivided 120 acres of land surrounding what would

Hollywood's spectacular advertisement for itself, the Hollywood sign is more solid than it looks: over 480 000 pounds of concrete, each letter 45 feet high. Erected in 1928, it fell into disrepair in the 70s but was reconstructed thanks to the contributions and fund-raising efforts of the Chamber of Commerce, Warner Brothers, Hugh Hefner, Alice Cooper, and other institutions and celebrities.

YWOOD

This page above There are over 600 independent film and television production companies in Los Angeles today, supporting a host of actors, costumers, prop companies and technical services. Hundreds of extras still phone Central Casting every afternoon hoping there's a job for them tomorrow.

Below As movie production and profits decline, studios looked for other ways to cash in on their assets. Some auctioned memorabilia, such as Dorothy's red slippers from *The Wizard of Oz*. Some began selling off their valuable real estate. 20th-Century Fox converted its back lot to Century City a multi-million dollar complex of office buildings, apartments, shops, restaurants and theaters.

Facing page Universal Studios have turned their 420 acre back lot, once a chicken ranch, into a popular tourist attraction. Visitors get a two hour tram tour of sets, special effects (including a flash flood, avalanche and the Red Sea parting), demonstrations by stunt men and make-up artists, and a trained animal show – a good overview of the inner workings of filmmaking.

DENTIST
CLOTHING
GENERAL STORE
GUNPOWDER
CHURCH & SONS LTD.

Paramount Pictures Corporation
VISITORS
ENTER

THE
BURBANK STUDIOS

become the intersection of Hollywood and Vine. Holly and wild oak were plentiful on the undeveloped acreage; hence, Mrs. Wilcox's inspiration.

In those days, watermelon and tomato crops were the basis of Hollywood's economy. But in 1903 there was no water with which to irrigate them, and the town was incorporated by Los Angeles, which had a more reliable water supply. Five years later, the fledgling film industry "discovered" Hollywood, and the legend was born.

The earliest movies had been made in the east; their makers went west for the cheap land, cheap labor, year-round sunlight and outdoor settings for Westerns. When D.W. Griffith arrived there in 1908, the population was still under 1000. A year later, there were 16 motion picture companies operating in Hollywood, and a tavern at the corner of Sunset Boulevard and Gower Street was turned into the town's first motion picture studio. The second, even more primitive, was converted from a barn, at the then wilderness of Sunset and Vine, by Cecil B. DeMille, for the Lasky Feature Players. He turned the horses' stalls into dressing rooms, rounded up extras off the streets and proceeded to direct *The Squaw Man,* the first full-length feature film made in Hollywood.

The Squaw Man was so successful that the studio rapidly expanded. Orange trees disappeared, replaced practically

Facing page above In 1927, the year that *The Jazz Singer* made silent movies obsolete, the major studios, Paramount, Universal, 20th Century Fox, Metro Goldwyn Mayer, Columbia, Warner Brothers and Republic – were already well established.

Below The fledgling film industry attracted the world's top directors, writers and actors all of whom built lavish mansions in Beverly Hills and weekend 'hideaways' at Malibu and Santa Monica. Today, the business is as unglamorous as this view of the Burbank Studios in the smoggy San Fernando Valley.

This page above A little girl is mesmerised by the hand prints and signature of Norma Shearer which are indelibly printed outside Grauman's Chinese Theater in Hollywood, along with the impressions of many other screen stars.

Left Here in the 'Walk of Fame' are the only stars to be seen on Hollywood Boulevard these days – 1500 of them, gold in color, each with the name of an entertainment personage in its center. Some of the names are obscure enough to stump the most dedicated film buff.

MANN'S
NESE
CKLE ROSE
LATE SHOWS FRI-SAT
THE MADISON
876 2006
Free
VALIDATED PARKING
FOR PATRONS OF MANN'S CHINESE THEATRES
HALF BLOCK SOUTH ON ORANGE DR.

overnight by new buildings and roads (although the original barn, subsequently moved to the Paramount back lot, still exists). In 1915, Griffith's *The Birth of a Nation* established the feature-length film as a legitimate art form.

In those days, with the stakes relatively low, making movies was nothing like the serious business it is today. It was exciting to experiment, take chances, make up the rules as you went along. For most people it was a game, played for fun and a bit of profit. Cameras were cranking away everywhere – on the beach, in the hills, in the center of town. But the tycoons who built the studios and dominated the industry for the next 50 years were immigrants, or sons of immigrants; they had worked hard since childhood, some under Dickensian conditions. William Fox had been in the dry cleaning business; Marcus Loew and Adolph Zukor were furriers; Lewis Selznick, a jewelry wholesaler; Sam Goldwyn, a glove salesman; Louis B. Mayer, a rag collector. What could they know, or care about fun?

During the 1920s, the studios imported talented men and women from all over the world. In a little more than two decades, the little town had become an international film center.

The movies' golden age lasted throughout the 20s and 30s and into the mid-40s, when weekly motion picture attendance in the United States peaked at 85 million. Then along came television, and the beginning of the end of the era of the major studios and the star system. When weekly attendance dropped to 17 300 000 in 1970, mergers and panic proliferated. Independent film and television production were formed. Low-budget films made by amateurs like Martin Scorsese, John Cassavetes and Mike Mazursky became more practical and foreign "runaway" production more profitable; the cheap labor that had once attracted moviemakers to Hollywood was now only available overseas.

Street vendors still hawk maps showing the alleged locations of movie stars' homes, but these tend to be

The forecourt of Mann's Chinese Theater, fomerly Grauman's Chinese, is still crowded with tourists inspecting the cement imprints of Barrymore's profile, Grable's leg and Bill Hart's gun, and comparing their hands and feet with those of other stars. Inside, the theater retains much of its original grandeur.

fictitious or historic documents, guides to former glories such as Pickfair (built by Douglas Fairbanks for his bride, Mary Pickford) and the ruins of the Houdini mansion on Mulholland Drive – and none of them are actually in Hollywood. But the most interesting features of Hollywood today aren't on any map. The eclectic clutter of architectural styles and shapes, for instance, constitute a mammoth outdoor art exhibit. Some commercial buildings have the shape of whatever is sold inside – hats, boats, hot dogs, pianos, donuts and the like. On the residential streets, L.A. retains her original, pre-"tract house" character. Stucco houses and apartment buildings painted every color of the rainbow are so close together that you can hear toilets flush in the next building; privacy is a low priority in Hollywood.

The Magic Castle, a private club for magicians, casts a wizardly spell with its Victorian turrets and towers on the hill behind Mann's Chinese Theater. Farther west on the hill, overlooking Sunset Boulevard, is the Chateau Marmont – the most charming apartment-hotel in town since the fabled Garden of Allah was knocked down to make way for a Savings and Loan office.

With the recent demise of Dolores's Drive-In, Barney's is the last refuge for characters who seem to be extras in a movie that has been in production for decades. They go there for the chili, for the apple pie, for the amazing assortment of music on the juke box, for turns at the pool table, and for auld lang syne.

The Farmer's Market, at Fairfax and Third Streets, attracts more tourists than locals now. Like Barney's, like the Chateau Marmont, the Farmer's Market is a piece of Hollywood's history. Nearby, however, is a piece of *pre*history: the La Brea Tar Pits. Forty thousand years ago, Ice Age creatures looking for water found tar, or *brea,* instead, and were trapped in it. When early Angelenos scooped up some of the tar to waterproof their roofs, they were astonished to find the fossilized skulls and bones of such long-extinct mammals as the sabre-toothed tiger, the imperial mammoth, the ground sloth and the American mastodon. More than a million specimens have been unearthed at this largest-known fossil site, and the finest are displayed in its museum.

The Hollywood Bowl's summer program includes dance companies, pop concerts and the L.A. Philharmonic's Symphonies Under the Stars.

Facing page In Disneyland's Frontierland, the triple-decker paddlewheel steamboat called *The Mark Twain* gives passengers a simulated tour of the rivers of America, past Tom Sawyer's Island, woodlands, Fort Wilderness and Indian territory. 30 years old and 75 acres large, Disneyland isn't the oldest or even the largest theme park in the country, but millions of satisfied customers would argue that it's the best – and certainly the cleanest!

This page left Even though Disneyland is in neighboring Orange Country, L.A. claims it – after all, it's only a 45-minute drive from downtown! The park is a triumph of imagination and technology, always changing but always retaining its charm for children of all ages.

Below The Mad Tea Party ride is one of the most popular at Disneyland. Like most of the theme park's attractions, it appeals to adults as much as to children. Also in Fantasyland are Mr Toad's Wild Ride, Snow White's Adventure, King Arthur's Carousel, Dumbo the Flying Elephant and Peter Pan's Flight.

HOLLYWOOD
Temple Hill Dr
6000 W

Facing page Hidden behind the trees lining innocuous-looking streets are houses in every conventional architectural style – Mexican, Spanish, French Chateau, English Tudor, American Colonial, Byzantine and Far Eastern – plus some startling freaks: a witch's cottage, an Egyptian temple, a miniature model castle with cannon on the battlements.

This page right If this street from the Old West looks familiar, it's because you've probably seen it in scores of movies made at Universal Studios, where this is now part of the tour that gives visitors a comprehensive look behind the scenes.

Below Hollywood overflows with artistic talent, much of it sadly underemployed. The artist who contributed these cheerful decorations to the Hollywood Community Center is anonymous, but all who pass by get a momentary lift from seeing his or her creations.

This page left J.R., alias Larry Hagman, does his shopping in exclusive Beverly Hills, as befits an oil tycoon or movie star.

Below Once the film-making center of the world, Hollywood today is the center of nothing. Even the Oscar ceremonies take place downtown, at the Music Center, and the annual international Filmex festival is held in Century City. The movie, television, music and video industries have fanned out into every section of Great Los Angeles, making her the *entertainment* capital of the world.

Facing page above If Hollywood is dead, Sylvester Stallone and his date (at the premier of *Staying Alive*) haven't heard the news.

Below The Sunset Strip, where billboards promise a new face, a new body, a new image, a new mate, even a new life.

The Los Angeles County Art Museum is so close to the tar pits that one can't help wondering whether it may share the fate of the fossils some day, assisted by the big earthquake that everyone talks about in terms of when, not if. Will the famous "Attack of the Killer Lipsticks" in the sculpture garden end up with the sabre-toothed tigers? Some art critics think it deserves to. A city as hooked on novelty as Los Angeles is bound to prefer modern to traditional art, but the collections of both are quite respectable. The most popular traveling exhibit ever displayed at the Museum was what some people referred to as the King Tut Show; Angelenos stood in line for hours and some scalpers got rich selling tickets to view the wonderful things in that most famous of tombs. Hollywood loved King Tutankhamun; whatever would the king have thought of Hollywood?

Wonderful things of a different kind are displayed in the boutiques on the west end of Melrose Avenue, but few of the prosperous-looking people who shop there actually live in Hollywood. Once a middle class, family neighborhood, Hollywood has become the domain of the would-be's, the has-beens and the never was's – the last place anyone would choose to raise a family. The Hollywood of the past, of the myths with which we all grew up, is gone forever – if, indeed, it ever existed. The Hollywood Memorial Cemetery, where Douglas Fairbanks, Rudolph Valentino, Tyrone Power and other idols rest, one hopes, in peace, is an appropriate tourist attraction. But the Hollywood sign still emblazons the hillside, a magnet to the continuing parade of new arrivals – starry-eyed hopefuls, looking for stardom or just new stars. As a symbol of their dreams, Hollywood will never die.

Major Attractions

1 Disneyland, the world's most famous theme park, now has 57 major attractions in its seven theme 'lands'. Tomorrowland, Frontierland, Fantasyland, Adventureland, Bear Country and Main Street. Fireworks displays on summer evenings.

2 Farmer's Market. More than 100 shops and restaurants sell and serve only the finest meat and produce. While you sip coffee or enjoy a meal at an outdoor café, you can watch an entertaining world of characters walk by.

3 Forest Lawn. The ultimate example of the American way of death. Original art and some excellent replicas are displayed in the Great Mausoleum, last resting place of many Hollywood celebrities.

4 The Getty Museum. Greek and Roman statuary, 18th-century French furniture and 14th-19th-century European paintings displayed in the late John Paul Getty's reproduction of a Roman villa.

5 Griffith Park. Four thousand acres containing hiking trails, golf course, outdoor theater, bird sanctuary, zoo, observatory, laserium, picnic areas and merry-go-round.

6 Hollywood Bowl. The natural amphitheater in the foothills is the Los Angeles Philharmonic's summer home. Pop concerts, ballet, an Easter Sunrise Service and special children's programs also take place at the Bowl.

7 Huntingdon Art Gallery, Library and Botanical Gardens. Fine art, rare manuscripts, and 200 lush acres of roses, camellias, desert plants and oaks. A world class garden, well worth the drive to San Marino.

8 Knott's Berry Farm. Once a ten-acre berry patch, now 150 acres of Wild West theme park with rides, ice shows, disco and live entertainment. Forty gift shops, 20 restaurants, and special celebrations such as Indian pow-wows and jamborees.

9 La Brea Tar Pits. The tar still bubbles around more than a thousand displays, including reconstructed models of the Paleolithic creatures whose fossilized remains were found in the pits.

10 Lion Country Safari. Lions, zebras, chimpanzees, giraffes, elephants, antelope and other wild African and South American animals roam freely around your car as you drive through this simulated African veldt.

11 Los Angeles County Art Museum. The three-building complex houses a permanent collection, traveling exhibits, a cafeteria, and a theater in which lectures, films and special programs are presented.

12 Mann's Theater, formerly Grauman's Chinese. Inside, it's a well-preserved relic of Hollywood's onetime glamor. Outside in the forecourt, impressions of the hands, feet and other outstanding features of the stars are on display in the concrete.

13 Marineland. Tanks the size of buildings contain exotic sea creatures that can be viewed from three levels. Outdoor shows star trained whales, sea lions and dolphins.

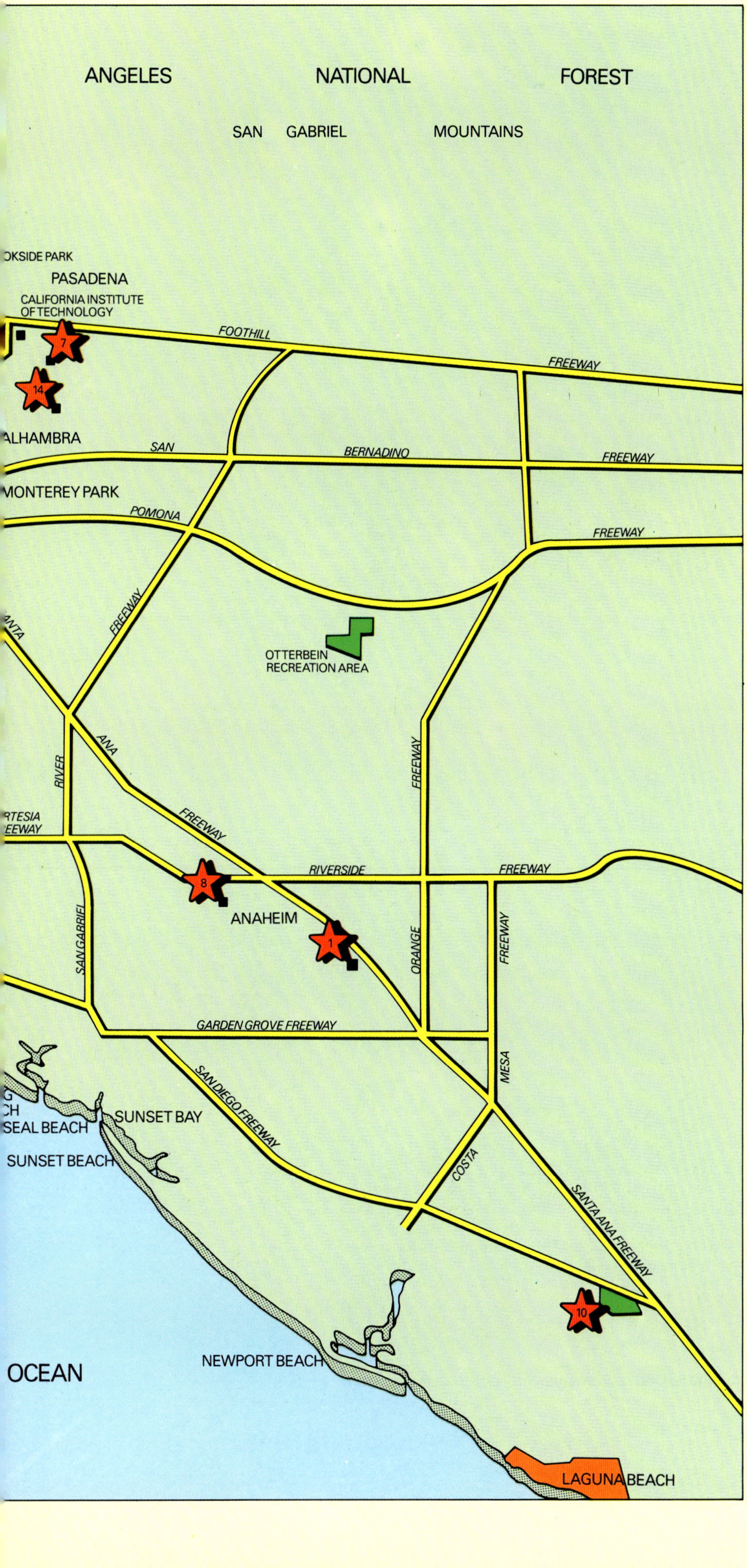

14 Missions San Fernando Rey de España and San Gabriel Arcangel. The seventeenth and fourth, respectively, in the chain of 21 missions built by the Spaniards more than two centuries ago. The padres worked, ate, slept, prayed and were buried here.

15 NBC Studios in beautiful downtown Burbank are the largest color studios in America. Visitors are given conducted tours of wardrobe, special effects, scenery and other behind-the-scene operations.

16 Queen Mary. One of the world's largest passenger liners, now moored at Long Beach and converted into a hotel, restaurant, convention center and Jacques Cousteau's 'Living Sea', a marine museum.

17 Universal Studios. Tours provide an excellent overview of the inner workings of making movies. There are special effects demonstrations, performances by stunt men and women, a trained animal show and much more.

18 Watts Towers. Simon Rodia's life work is an example of eccentric, unique folk art that must be seen to be believed.

19Westwood Village. One of the best places in L.A. to walk, window-shop and watch the people. Interesting shops, excellent restaurants, lots of first-run movie theaters and beautiful people, many of them from the nearby UCLA campus.

20 Venice Boardwalk. A comprehensive view of L.A.'s beach life – roller skaters, body builders, bicyclers, joggers and surfers put on an entertaining free show every weekend of the year.

PICTURE CREDITS

All Sport 18, 38 bottom, 42 bottom, 43 top **Camerapix Hutchinson** 9 bottom, 35, 39 bottom **Bruce Coleman Limited** 8 bottom, 20 top, 25 top, 28 top, 45 top **Colorific!** 19, 34 bottom, 51 top **Daily Telegraph** 11 bottom, 27 bottom, 34 top, 38 top left, 59 top, 65 bottom **The Image Bank** 13, 27 top, 58 bottom **Network/Barry Lewis** 14 bottom, 31, 38 top right, 39 top, 40, 46 center, 47 top, 50 bottom, 51 bottom, 59 bottom, 69 bottom **The Photographers Library** 1, 10, 28 bottom, 30 top, 48-49, 52, 67 top **The Photo Source** 21, 26, 32-33, 36-37, 53 bottom, 64, 72 **Rex Features** 42 top, 43 bottom **Frank Spooner Pictures** 41, 45 bottom, 58 top, 68, 69 **Tony Stone Associates** front & back covers, 4-5, 6-7, 15 right, 22-23, 47 bottom, 60-61, 62-63, 65 top **Vautier-de Nanxe** 8 top, 14 top, 15 left, 24, 25 bottom, 29 bottom, 44, 46 top & bottom, 56 **Vision International** 2-3, 11 top, 29 top, 50 top, 53 top, 57, 67 bottom **ZEFA** end papers, 9 top, 12 bottom, 16-17, 20 bottom, 23, 30 bottom, 54-55, 66

Multimedia Publications (UK) Limited have endeavored to observe the legal requirements with regard to the rights of the suppliers of photographic material.

BANK OF AMERICA